Fire Safe Student Housing:
A Guide for Campus Housing Administrators

Frederick W. Mowrer, Ph.D., P.E.
Department of Fire Protection Engineering
University of Maryland
College Park, Maryland 20742

February 1, 1999

Table of Contents

Disclaimer

References to websites of manufacturers of fire protection equipment are not intended as endorsements for the products of the manufacturers cited. They are meant only to serve as examples of resources available for further information regarding different types of fire protection equipment.

Introduction

As a campus or fraternal housing administrator, you have the duty to provide a safe environment for the college students housed in your facilities. You are faced with a number of different issues competing for your attention and for the limited resources available to address them. You must decide how to best allocate available resources among competing demands and interests. To make these decisions wisely, you need to understand the risk factors involved, the alternatives available to you and the relative costs and benefits of the different options.

Fire is one of the deadlier perils that threaten student safety. A fire in a student housing facility can quickly rage out of control if appropriate safeguards are not in place to stop it. While fatal fires in student housing are not an every day occurrence, they can and do happen, perhaps more frequently than you recognize. These fires do not make the headlines unless a number of students are killed, so it is easy to underestimate the risk of fire in student housing facilities. It is important for you to recognize that fire safe student housing does not just happen by chance, nor can it be taken for granted. It requires an ongoing commitment on the part of the community, the institution and the administration. Careful planning, implementation and maintenance are all essential ingredients of a successful fire safety program for student housing.

The purpose of this report and the accompanying video is to provide campus housing administrators, fraternal organizations and others responsible for housing college students with an overview of the elements of fire safe student housing. The goal is to present a balanced approach that will permit housing administrators to make risk-informed decisions regarding the costs and benefits associated with different fire safety features and levels of fire protection. Additional resources that are available to help in the development of a comprehensive program for fire safe student housing are also identified.

Ultimately, student-housing administrators need to seriously consider the installation of automatic sprinkler systems in the residential facilities they manage. These systems have an established record of preventing catastrophic fires in residential facilities, making sprinkler protection perhaps the single most effective weapon in the residential building fire safety arsenal. Over the past 15 years, the hospitality industry in the United States has embarked on an ambitious program to install sprinkler protection in most hotels and motels. As part of this effort, various technologies have been developed to reduce the costs, aesthetic impacts and inconveniences associated with the installation of automatic sprinklers in existing residential facilities. These technologies translate directly to both new and existing student housing facilities, providing the opportunity for college students to enjoy the same high level of fire protection as the traveling public.

Background

On any one campus or in any one college community, building fires are relatively rare events. As a result, it is easy, perhaps even natural, to become complacent about fire safety, to confuse good luck with good practice. But when fires do occur, and they do, they can develop with incredible speed and have devastating consequences. When deaths and disfiguring burn injuries result, the consequences last forever, impacting on not only the victims and their families, but on the entire college community as well. For example, an arson fire that killed two students in a dormitory at Ohio State University in 1968 is still remembered – 30 years later – as perhaps the worst tragedy to ever strike the OSU campus community.

A fire in a fraternity house at the University of North Carolina on Mother's Day in 1996 further illustrates the type of devastation that fires on campus can wreak. Following a celebration during the spring graduation weekend, a fire developed in the basement recreation room of the fraternity during the wee hours of the morning. Fed by the combustible interior finish and furnishings, the fire reached hazardous proportions while residents slept. The fire swept through the structure, leaving five students dead and one student, the only survivor, injured in its wake. What had started as an annual spring celebration ended with a somber memorial service.

The fire at the University of North Carolina is just one example of campus housing fires. Table 1 presents a summary of a review of student housing fires that received news media attention during the 20-year period from 1979 to 1998. Reports of these fires were obtained primarily from the Lexis-Nexis® Academic Universe website. Unfortunately, these reports tend to be preliminary and sketchy. Nonetheless, as indicated in Table 1, multiple death fire scenarios are not very common in student housing; most fatalities occur by ones or twos. This is similar to the general population, where most fire fatalities occur by ones or twos in private residences. Table 1 also indicates, however, that fires in campus housing can displace many students at one time, creating a logistical problem during the middle of a school term.

Dr. John L. Bryan, Professor Emeritus in the Department of Fire Protection Engineering at the University of Maryland, recently completed a detailed examination of selected college dormitory and fraternity house fires in connection with this project. Bryan selected fifteen fires, including nine dormitory and six fraternity house fires, from 1967 to 1996 for detailed analysis. These fire incidents were selected based on the occurrence of fatalities or injuries to occupants along with the availability of a published report for each incident. These fifteen fire incidents resulted in 44 reported fatalities and 143 reported injuries. Bryan's comprehensive report is attached as Appendix A to this report for reference.

Table 1. Student housing fires from 1979 to 1998 that received news media attention.

University	Date of Fire	Housing Type	Cause of Fire	Property Loss ($)	# Students Displaced	Fatalities/ Injuries
Nebraska Wesleyan Univ.	Oct. 8, 1998	Off-Campus Apartment	Cigarette	50, 000	6	1 / 0
Alfred U. (NY)	Oct. 10, 1998	Dormitory	Light bulb	NA	140	0 / 0
Kalamazoo College (MI)	Sept. 18, 1998	Dormitory	Arson	NA	NA	0 / 0
Murray State (KY)	Sept. 18, 1998	Dormitory	Arson	NA	+100	1 / 16
Ohio State Univ.	Sept. 2, 1998	Off-Campus Apartment	Arson	28,000	+4	0 / 0
Univ. of Buffalo (NY)	July 26, 1998	Dormitory	Unknown	100,000	25	0 / 0
Univ. of Arizona	July 14, 1998	Fraternity	Arson	NA	39	0 / 0
Greenville College (IL)	Dec. 9, 1997	Dormitory	Unknown	NA	40	1 / 7
Johns Hopkins U. (MD)	Aug. 31, 1997	Off-Campus House	Cigarette	NA	+3	1 / 0
Lindenwood College (IL)	April 17, 1997	Dormitory	Electrical	10,000	NA	0 / 0
School of Visual Arts (NY)	Feb. 21, 1997	Dormitory	Cigarette	NA	+50	1 / 0
U of C Berkeley (CA)	Jan. 9, 1997	Fraternity	Candle	NA	+15	0 / 2
Central Missouri St. Univ.	Jan. 3, 1997	Dormitory	Arson	NA	NA	1 / 0
Ohio Wesleyan U. (OH)	Oct. 19, 1997	Fraternity	Unknown	NA	NA	1 / 0
Ohio State U. (OH)	Aug. 13, 1996	Off-Campus Apartment	Electrical	20,000	1	0 / 0
William Jewell College (KS)	Aug. 8, 1996	Fraternity	Cigarette	500,000	+5	0 / 0
Univ. of N. Carolina	May 12, 1996	Fraternity	Cigarette	NA	+10	5 / 3
Mesa State (CO)	Dec. 21, 1995	Off-Campus Apartment	Unknown	NA	3	1 / 3

University	Date of Fire	Housing Type	Cause of Fire	Property Loss ($)	# Students Displaced	Fatalities/ Injuries
Marshall College (PA)	Aug. 5, 1995	Fraternity	Arson	NA	NA	0 / 0
Univ. of Florida	March 10, 1995	Fraternity	Unknown	NA	+6	0 / 0
Ohio State Univ	Nov. 22, 1994	Fraternity	Suspicious	+20,000	+25	0 / 0
U of C Berkeley (CA)	Aug. 15, 1994	Fraternity	Unknown	200,000	25	0 / 0
Univ. of Wisconsin	Oct. 26, 1993	Sorority	Unknown	+100,000	10	1 / 2
Ohio State Univ	May 1996	Fraternity	Unknown	500,000	NA	0 / 1
Drexel Univ. (PA)	Feb. 18, 1993	Fraternity	Unknown	NA	NA	0/ 1
State Univ at Stony Brook	Feb. 25, 1992	Dormitory	Unknown	NA	200	0 / 1
San Jose State (CA)	Oct. 19, 1990	Dormitory	Unknown	NA	178	0 / 20
U of C Berkeley (CA)	Sept. 1990	Fraternity	Unknown	NA	NA	3 / 0
Univ. of Washington	July 20, 1990	Sorority	Spontaneous Combustion	130,000	NA	0 / 0
Rutgers Univ. (NJ) *	July 18, 1990	Fraternity	Arson	NA	NA	0 / 0
Northern Illinois Univ. (IL)	Feb. 25, 1989	Dormitory	Suspicious	+1,000	+50	0 / 0
Univ. of Mississippi	Aug. 4, 1988	Fraternity	Suspicious	100,000	NA	0 / 0
Columbia Univ.	Jan. 1, 1987	Fraternity	Electrical	NA	+5	0 / 5
Univ. of South Carolina	Sept. 5, 1986	Fraternity	Electrical	450,000	58	0 / 3
Indiana Univ.	Oct.22, 1984	Fraternity	Arson	NA	+30	1 / +30
George Washington Univ. (DC)	April 20, 1979	Dormitory	Unknown	NA	+35	0 / 35

*Represents a string of arson fires that occurred simultaneously in three fraternity houses on the campus. Fortunately there were no injuries reported from the incident.

Bryan analyzed a number of variables associated with these incidents, including ignition and propagation variables, construction variables, occupant behavior variables and fire protection system variables. He also analyzed two sets of data compiled by the National Fire Protection Association regarding dormitory, fraternity and sorority fires. The first set of data, published in 1955, was based on fires during the period from 1944 to 1954. The second set, published in 1995, covered the period from 1990 through 1994. Bryan notes the significant social and cultural differences in the campus environment between these two surveys, particularly changes in the supervision of residential facilities and the restrictions placed on student residents.

Some disconcerting trends arise from the data. Bryan notes that the occupant behavior activities of incendiary fire setting, cooking and smoking appear to be the primary causes of student housing fires, with alcohol consumption being a significant factor. Most troublesome is the increase in the incidence of incendiary and suspicious fires between the first and second data sets. Such causes constituted about 10 percent of the fires in the 1955 data, but jumped to almost 20 percent of the fraternity and sorority fires and 30 percent of the dormitory fires in the 1995 data. In 1955, incendiary or suspicious fires ranked fifth as a causative factor; in 1995, incendiary or suspicious fires ranked first. While arson can never be condoned, neither can it be ignored when it comes to fire safe student housing.

Bryan further notes the significant role of highly combustible upholstered furniture in the student housing fires he analyzed. Upholstered furniture, predominately sofas, were the fuel material ignited first in seven of the fifteen fires he analyzed. Because of this, Bryan concludes that procedures should be initiated to regulate the inclusion of new highly combustible upholstered furniture into dormitories, fraternities and sororities. Based on his analysis, Bryan also concludes that procedures should be initiated to provide for the installation of smoke alarms in student rooms and automatic sprinklers throughout new dormitories, fraternities and sororities, as well as in existing facilities when they are renovated.

Fatal fires are always difficult to accept; when they occur in student housing, they are particularly devastating. There are a number of reasons for this. Most college students, particularly those in campus or fraternal housing, are living away from the security of their parents' homes for the first time. Parents, sending their children off to college, do so with a mixture of pride and trepidation, but certainly with the expectation that the college community will provide a reasonably safe environment for their loved ones.

On the part of the students, a certain sense of immortality seems to come with the territory as they embark on this exciting period of independence. Many students do not yet have the maturity or experience to recognize real threats to their personal safety; consequently, they sometimes indulge in foolish, even dangerous, behavior without realizing the risks or potential consequences. When it comes to fire safety, most students are uneducated; that is, unless they have been properly trained in fire prevention and response should a fire occur.

Because of the relatively rare occurrence of building fires, few people outside the fire profession have the experience or knowledge to appreciate sometimes subtle differences between fire safe structures and those that will become hazardous when a fire occurs. Fewer still, even among fire professionals, fully appreciate the incredible speed with which fires can develop in buildings or how quickly escape routes can be blocked if appropriate fire safety features are not present or are compromised. **Accidental fires in residential facilities can reach deadly proportions in less than three minutes after ignition, incendiary fires even faster!**

As a campus or fraternal housing administrator, you are probably aware of the local and state fire safety regulations that apply to your student housing facilities. These regulations impose specific minimum requirements with respect to the building fire safety features required by law. What you may not know is that the Hotel and Motel Fire Safety Act of 1990 (PL101-391) also applies to your campus if it is used for federally funded meetings and conferences.

The Hotel and Motel Fire Safety Act of 1990 (PL101-391) was passed into law by Congress to save lives and protect property by promoting fire and life safety in hotels, motels and other places of public accommodation. The law encourages and eventually mandates that federal employees on travel must stay in public accommodations that adhere to the life safety requirements in the legislation guidelines. PL101-391 also states that federally funded meetings and conferences cannot be held in properties that do not comply with the law.

PL101-391 is applicable to all places of public accommodation, and requires that such properties are equipped with:

- hard-wired, single-station smoke detectors in each guestroom in accordance with the National Fire Protection Association (NFPA) standard 72;

- an automatic sprinkler system, with a sprinkler head in each guest room in compliance with NFPA standards 13 or 13R.

Properties three stories or lower in height are exempt from the sprinkler requirement.

Realistically, it can be difficult to obtain the resources to install fire protection systems if prevailing regulations do not require such systems. Therefore, if existing state and local regulations or the federal Hotel and Motel Fire Safety Act do not provide sufficient incentive, it may be necessary to pursue the local adoption of regulations requiring such systems. A number of communities have already instituted regulations requiring the installation of automatic sprinkler systems in college housing facilities. Many of these ordinances have been adopted in response to local tragedies, but the lessons learned should not be restricted to any one campus or community.

A number of resources are available to aid in the development and implementation of local ordinances for sprinkler protection. Some of these can be obtained at the following websites:

- www.nfpa.org
- www.nfsa.org
- www.firesprinkler.org

Many of the concepts discussed here are the same in principle as those contained in nationally recognized standards, but specific standards adopted by law should be consulted to assure at least a minimum level of regulatory compliance.

The Elements of Fire Safe Student Housing

The fire safety of student housing can be considered in terms of four primary elements:

- Prevention
- Occupant awareness and training
- Detection and alarm
- Suppression

Together, these four elements have the acronym "PODS." The PODS acronym is appropriate for student housing because the term connotes a protected living environment. The elements of the PODS concept are discussed.

Prevention

The first element of the PODS concept is prevention. Fires require three elements to occur: fuel, air and an ignition source. These three elements have traditionally been illustrated in terms of the "Fire Triangle" to show the relationship between elements. The prevention of fire requires that one or more of the elements of the fire triangle be removed. Since air is always present and available in the atmosphere, the prevention of fire generally requires control or elimination of either fuels or ignition sources, or separation of the two through appropriate safeguards.

Potential fuels present in student housing take a myriad of forms, including:

- Upholstered furniture, mattresses and bedding
- Draperies, curtains and other free-hanging decorations
- Combustible wall, ceiling and floor finishes
- Desks, dressers and bookcases
- Books, papers, notebooks and reports
- Trash and recycling materials
- Clothing
- Stored commodities

Upholstered furniture has been implicated in many serious fires in student housing facilities. Today, these products are typically padded with polyurethane foam. Once ignited, these products can burn with incredible speed and intensity. As a consequence of these hazardous burning characteristics, many authorities now require upholstered furniture intended for commercial or institutional usage to be made of fire retardant materials and assemblies. Such products can greatly reduce the potential for the rapid fire development and hazardous conditions associated with non-fire retardant upholstered furniture. As noted by Bryan, consideration should be given to the specification of fire retardant upholstered furniture and mattresses for use in campus housing even if local regulations do not mandate such products.

Mattresses, like upholstered furniture, are also typically padded with polyurethane foam. Once ignited, mattresses and bedding can also burn with incredible speed and intensity. Bunk beds, which are fairly popular in student housing, compound the problem because of the vertical stacking of the mattresses. While not traditionally required in residential facilities, fire retardant mattresses have been developed for institutional occupancies. Their usage is becoming more widespread. Even with fire retardant mattresses, however, the flammability of the bedding materials needs to be considered.

Draperies, curtains and wall decorations, such as posters, represent thin fuels that can be relatively easy to ignite with small ignition sources, such as candles. Because of their relative ease of ignition and vertical orientation, these products can rapidly spread a fire beyond ready control. As a consequence of this potential for ignition and fire development, some regulations require draperies, curtains and other free-hanging decorations to be fire retardant. For example, the Life Safety Code

(NFPA 101), a widely adopted standard, requires these products to be fire retardant in new and existing hotels and dormitories. Even if not mandated by law, regulation of the flammability of these products would make a difference for some fire scenarios. It would also require diligence on the part of the housing administrator to assure that only approved draperies, curtains and wall coverings were used.

Wall, ceiling and floor finishes have a broad range of flammability characteristics. By virtue of the large areas covered by these finishes, combustible finishes all pose some risk of fire spread. The degree of this threat depends on a number of factors, including the flammability characteristics of the finish and the fire conditions to which it is exposed. Combustible wall and ceiling finishes have been implicated in the development of a number of serious fires, including the 1996 Mother's Day fire at the

University of North Carolina. Because of their location on the floor, combustible floor finishes are less likely to contribute to the early development of a fire. However, these same products can contribute significantly to a fully-developed fire. If safeguards, such as automatic sprinkler protection, are not in place to prevent fully-developed fire conditions, consideration should be given to regulating the flammability characteristics of floor coverings. Carpets and other soft floor coverings should never be used on walls or ceilings unless they have been specifically qualified for these applications by fire test.

Paper products, including books, notebooks, trash and recycling, are common fuels in student housing facilities. These products will burn with a variety of intensities, depending in large part on the quantity and whether the paper is loosely or tightly packed. Trash and recycling paper, and the containers in which they are deposited, pose a particularly significant risk because the paper products are more likely to be loosely packed and there is a greater potential for ignition, from improperly discarded smoking materials for example. Trash receptacles have been implicated in enough fires that a simulated trash receptacle fire is used as the standard ignition source in a number of fire test methods. One way to reduce the incidence and intensity of trash and recycling fires would be through the use of covered metal or fire retardant waste and recycling containers. Such receptacles are currently required in some applications.

Desks, dressers, bookcases and other casework have traditionally been made of fairly dense wood-based materials. Because of their high density, these products are more difficult to ignite than padded furniture or decorations, thus affording a higher degree of fire resistance than these other products. However, once ignited, these wood-based products can contribute a significant amount of heat to a developing fire, so their potential for contribution to a fire should not be ignored. Casework made of materials other than wood or steel should be carefully evaluated for its ease of ignition and potential contribution to fire development.

When not strewn about a student's room, clothing is usually stored in a dresser or in a closet or wardrobe. Clothing stored in a closed dresser, closet or wardrobe is generally protected from early involvement in a developing fire. But fires that initiate in open closets or wardrobes can develop with surprising speed due to the types of fuel and confined geometries involved.

Finally, stored commodities can run the gamut from innocuous to extremely hazardous. Particular care should be taken to assure that flammable gases and liquids are not stored in residential buildings without appropriate safeguards. Everyday products such as lawnmowers and gas-fired barbecue or camp grills are easy to overlook, but can represent significant sources of easy-to-ignite fuel if accidentally released. While not common in dormitories, such products are more likely to be found in fraternities, sororities and off-campus housing.

Potential ignition sources present in student housing also take a myriad of forms, including:

- Smoking materials, including cigarettes, matches and lighters
- Candles and incense
- Cooking equipment and appliances
- Electric lamps and appliances
- Building services, including electrical and gas distribution and utilization equipment
- Arson and other incendiary or suspicious devices.

Careless disposal of smoking materials is the sixth ranked cause of reported residential fires, but is the leading ignition source for residential fire deaths. This difference in relative ranking can be attributed to two factors. Of primary importance, people sometimes fall asleep or pass out while smoking. When this happens, the smoker is frequently a victim if a fire develops because of the proximity to the fire. Secondly, fires ignited by smoking materials can smolder for extended periods of time before developing into hazardous fires. Such fires are more likely to go undiscovered than fires that develop immediately upon ignition. In the meantime, residents may go to sleep and fail to discover such a fire until roused by an alarm system or some other indicator. Depending on the nature of the fire, its location relative to the sleeping residents and the fire protection features in place, the time for escape between notification and the development of hazardous conditions may be very short, on the order of a few minutes or even less.

Smoking in campus housing should be vigorously regulated, if not prohibited. Most public buildings, including college housing facilities, have become smoke-free in recent years, but a sudden ban on smoking may be counterproductive by forcing smokers into the closet, both figuratively and literally. A better alternative may be to provide designated smoking areas that are properly equipped with suitable ashtrays and other fire protection features.

Candles and incense are particularly popular in college residences. The use of such devices should be regulated through appropriate policies, if not prohibited outright as on many campuses. Students should be instructed to use appropriate candle and incense holders, to locate such holders away from other combustibles, such as draperies, and to never leave such devices burning unattended. Violations of such policies should be considered serious breaches of student housing rules.

Cooking is the leading cause of all residential fires, but only the sixth leading cause of fire deaths. The incidence of cooking fires in college dormitories will be lower because such housing units do not typically have kitchens. Nonetheless, such units frequently have cooking appliances like hotplates, coffee makers and hot water immersion heaters. In some respects, such devices pose an even greater fire risk because they are being installed in spaces not designed for such usage. Other campus housing units, such as apartments or group homes, do have kitchens and thus are prone to the same types of kitchen fires as residences in general. Such fires frequently result from unattended cooking or from wearing loose-fitting apparel when cooking.

Electric lamps and other electric appliances generate heat. If this heat is not adequately dissipated, a fire can occur. Lamps can cause fires in a number of ways, including:

- If a lamp is covered with a combustible material, such as an article of clothing, a towel or some other fabric, the fabric can ignite;
- If a lamp tips over, the bulb can come in contact with a combustible material, such as bedding or upholstery, and cause it to ignite;
- A bulb might explode, perhaps as a result of a lamp tipping over, permitting the hot filament to come in contact with an easily ignitable material.

Torchiere lamps with halogen bulbs pose a particularly severe fire hazard. These halogen bulbs become much hotter than traditional incandescent bulbs. Recently, the Consumer Product Safety Commission has addressed the fire safety of these devices as a result of a number of fires caused by them. The CPSC has required manufacturers to distribute a wire guard for existing lamps to prevent fabrics from coming in contact with the halogen bulb. New lamps have an improved design that incorporates such a guard. Nonetheless, there are many such lamps without guards. Even with guards, there are scenarios where such lamps can cause ignition. Some campuses have banned the use of torchiere lamps in residential facilities.

Electric appliances, including radios, stereos, televisions and computers, require adequate ventilation to prevent heat buildup. If the ventilation to such devices is restricted, overheating to the point of ignition is possible. Some devices have thermally-activated safety switches to prevent runaway overheating, but others do not. Despite the safety features in place, wherever electricity is utilized there is a possibility of a short circuit or other malfunction that could cause ignition under the right circumstances.

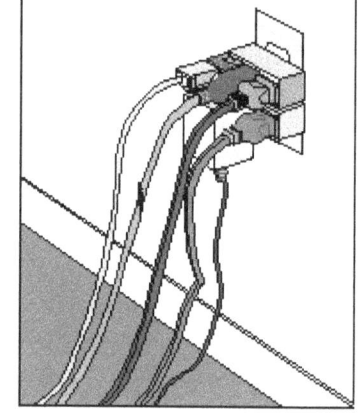

Building services, including the electrical distribution system and gas appliances, also present some potential to act as ignition sources. The installation of such building services is generally regulated, but older buildings may have substandard installations relative to current standards and usage. In particular, the electrical service provided in older buildings may not have the capacity for the range of modern electrical equipment now in everyday use. Without upgrade, such systems may be overloaded, with students resorting to the use of extension cords and power splitters to get the power they need where they need it. Such practices increase the potential for fire.

Finally, incendiary and suspicious fires constitute the third leading cause of all residential fires, the second leading cause of fire deaths and the leading cause of property loss. Incendiary and suspicious fires have been identified as the leading cause of campus housing fires, constituting between 20 and 30 percent of all fire causes in such facilities. While such intentional fires are difficult to prevent completely, other fire protection measures can be used to reduce the potential for and impact of arson fires.

This fairly lengthy discussion of potential fuels and ignition sources should make it clear that complete elimination or control of potential fuels and ignition sources is not practical. Other safeguards are also needed to provide a reasonable level of fire safety. This does not mean that efforts to control the flammability characteristics of potential fuels and the presence and safe use of potential ignition sources are for naught. While such efforts may not be completely effective, they can contribute to a significant reduction in fire incidents and losses. Unfortunately, measuring this reduction is difficult, if not impossible, particularly in the short term. While it is possible to measure changes in the numbers of fires and associated losses over time as a gauge of fire prevention effectiveness, it is hard to know how many fires may have been prevented by a particular strategy.

Most campuses have prepared fire safety standards and regulations. Many campuses now publish these standards online as a service to the campus community. Some of these websites are identified below. These and other resources useful for identifying the elements and implementation of a fire prevention plan include:

- www.usfa.fema.gov/safety/
- www.pp.okstate.edu/ehs/
- www.inform.umd.edu/CampusInfo/Departments/EnvirSafety/fire/fire/campus.html
- www.inform.umd.edu/CampusInfo/Departments/EnvirSafety/fire/greek/index.html
- www.cco.caltech.edu/~safety/fpp.html
- http://www-ehs.ucsd.edu/fire.htm
- www.rci.rutgers.edu/~zuccare/reshall.html
- www.ou.edu/oupd/fireprim.htm

Occupant awareness and training

The second element of the PODS concept is occupant awareness and training. There are two aspects to this element:

- Fire prevention training
- Fire response training

Fire prevention training should include instruction on what can be done to reduce the potential for ignition. The issues discussed previously with respect to fire prevention need to be conveyed to students to make them more fire aware. Students should be trained to recognize potentially hazardous situations, such as smoking in bed, careless use of candles and cooking, use of excessive flammable decorations, poor housekeeping practices and blockage of exit paths.

Fire response training should include clear instruction on what residents should do in the event of a fire. The first decision a resident must make is whether to fight the fire, notify other residents or evacuate immediately. Because of the rapidly changing nature of many building fires, this decision will depend in large part on the location and state of the fire when it is discovered, which in turn will depend on the fire safety features of the building. Relatively small fires can be fought successfully with portable fire extinguishers. If residents are expected to fight small fires, however, they need to be trained in the location and operation of portable fire extinguishers, as well as how to recognize when a fire is too large to attack.

Evacuation behavior should be rehearsed. Residents should know at least two ways out of the building, a primary path and a secondary path. Paths should be checked for safety before proceeding; doors should be felt for heat before opening them, then they should be carefully cracked open to check conditions on the other side. People should stay low, where the air is generally less smoky, and proceed with deliberate speed to the exit and out the building. Once outside, residents should not reenter the building for clothing, valuables or pets. These rules for safe evacuation behavior should be practiced periodically.

Whether they are going to fight or flee, residents need to understand how dangerous fires behave. This includes an appreciation of how fast building fires can develop and familiarity with the phenomenon of *flashover*. Flashover is a brief period of transition in a fire when virtually every combustible in a room ignites and the room becomes engulfed in fire. Conditions within the room cannot be survived

beyond flashover and conditions in other parts of the building will deteriorate rapidly as large quantities of heat, smoke and carbon monoxide are pumped from the fire room after flashover. Without intervention, the time from ignition to flashover can be as brief as a few minutes in many realistic fire scenarios.

Residents also need to understand the fire safety features of the building. Do rooms have self-closing doors? What is the purpose of the self-closing feature? How are the exits arranged? What are the primary and secondary egress paths? What should be done if the egress paths cannot be used? Does the building have a fire detection and alarm system? If so, how is the alarm system activated and what does the fire alarm sound like? Does the alarm system automatically notify the fire department or must the fire department be notified by phone? Does the building have an automatic sprinkler system? What does the sprinkler system alarm sound like? These features and issues need to be discussed with residents in a way that educates without overwhelming.

Residents should be instructed not to tamper with the building fire protection features, either intentionally or accidentally. Smoke detectors should not be covered or have the batteries removed. Sprinklers should not be heated or subjected to physical abuse, such as from thrown articles like frisbees or footballs. Objects should not be stored in exitways, including hallways, corridors and stairways, where they could obstruct the egress path or serve as potential fuels for a fire. Doors to rooms and exits should not be blocked open. In the event of a fire, doors should be shut to hinder the progress of the fire and the spread of smoke. Severe sanctions should be imposed on students who violate these rules.

The resources identified at the end of the Prevention section also contain valuable information regarding occupant awareness and training. In addition, the Eau Claire Fire Department has produced a videotape and accompanying literature under US Fire Administration sponsorship. This videotape, entitled "Get Out – Stay Out," is intended for students to view. Contact the USFA (www.usfa.fema.gov) to obtain this information package.

Detection and Alarm

The third element of the PODS concept is detection and alarm. Automatic fire detection is a key fire safety feature in any residential building, from detached single family houses to modern high rise apartments or dormitories. In particular, smoke alarms installed throughout a building permit early detection and notification of incipient fires, particularly while residents sleep. Different fire safety regulations have somewhat different requirements with respect to fire detection and alarm systems.

There are three basic types of fire detection and alarm systems suitable for use in residential facilities. These include:

- Single- and multiple-station smoke alarms
- Zoned fire detection and alarm systems
- Addressable fire detection and alarm systems

An approved, single-station smoke alarm should be installed in every sleeping room as well as in every living area within a suite of rooms. At least one smoke alarm should be provided on each level of multi-level units. Nationally recognized fire safety codes and standards typically require this level of protection at a minimum.

Battery-operated single-station smoke alarms are generally permitted in existing buildings, but it must be recognized that these devices require regular battery replacement, typically at least annually. Furthermore, the batteries used in these devices are usually the 9-volt batteries used in many consumer electronics, so "borrowing" of smoke alarm batteries is a fairly common problem in student housing. These factors reduce the reliability of battery-operated smoke alarms

and demand a relatively high level of ongoing maintenance. A survey sponsored by the Consumer Product Safety Commission determined that approximately one-third of battery-operated smoke alarms are not operational; this can leave residents with inadequate protection and a false sense of security.

A better alternative to battery-operated devices is "hard-wired" single-station smoke alarms. These devices receive their primary power from a 120-volt electrical circuit in the building. This arrangement is typically required in new buildings, where it is a relatively simple matter to provide electric power to the smoke alarms. The Hotel and Motel Fire Safety Act of

1990 also requires "hard-wired" smoke alarms, even in existing buildings. An even higher level of protection is achieved if hard-wired smoke alarms are provided with a battery as a secondary power supply; such devices will continue to operate in the event of a power outage, which might be caused by a fire itself. As these pictures show, however, it may be difficult for an untrained person to tell the difference between a battery-operated and a hard-wired smoke detector. You should be aware of the types of devices installed in your facilities and of the maintenance issues related to their reliable operation.

In new construction, smoke alarms within a living unit or suite of rooms typically must be arranged so that operation of any smoke alarm within the living unit will cause all smoke alarms within the living unit to sound. This feature is valuable in large living units, whether new or existing, because a smoke alarm operating in a remote part of the unit may not be audible in the sleeping areas. This multi-station arrangement does not extend to smoke alarms in individual dormitory or fraternity rooms, but would apply to multi-room suites with common living areas.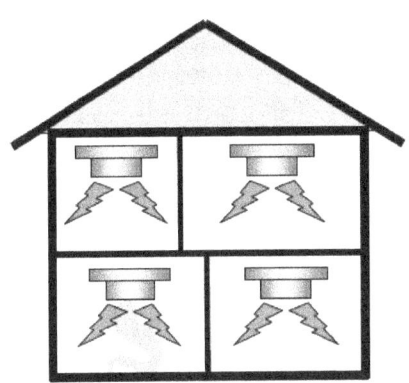

The purpose of single- and multiple-station smoke alarms is to notify people in the immediate living unit of the alarm condition. They can also be used to provide remote annunciation of an alarm condition at a monitored location, although this is not generally required. Single- and multiple-station smoke alarms are not intended to actuate the building fire alarm system; some nationally recognized codes actually prohibit actuation of the building fire alarm system by these devices. This prohibition is due to the relatively high incidence of nuisance alarms associated with smoke detection systems.

Historically, smoke detection systems have produced more nuisance alarms than real alarms. Various estimates suggest that for every actual alarm there have been 10 to 20 nuisance alarms. These nuisance alarms are typically caused by non-fire sources, such as cooking or steam from a bathroom, that are misinterpreted as fire signatures. Many of these nuisance alarms can be avoided through appropriate system design. Regardless of the cause or remedy, however, this "crying wolf" syndrome has caused many people to routinely ignore fire alarms in buildings without independent confirmation of actual fire signatures, such as the smell of smoke. In large buildings, such independent confirmation may not occur until the fire has developed to a hazardous point. In the meantime, valuable time is wasted.

The technology of fire detection and alarm systems has advanced in recent years to reduce the incidence of nuisance alarms. Where old technology fire detectors were either on or off, the sensitivity of new, analog detectors can be monitored and adjusted. These detectors can even identify when they need to be cleaned. Where old alarm systems were typically zoned to indicate the general area of detector operation, new systems are addressable so they can pinpoint the specific location of an operating detector. These two features tremendously increase the reliability of fire detection and alarm systems, particularly for difficult applications like dormitories. Now, for example, if a detector activates in a particular room, the room can be identified immediately and contacted by phone to determine if the source of the smoke is an actual fire or a nuisance source. If a fire is confirmed, or if there is no answer to the call, appropriate actions can be initiated. New detectors can also be programmed to recognize certain signal patterns from nuisance sources as being distinct from fire smoke. This discrimination further increases the reliability of such systems.

Costs for fire detection and alarm systems and components range from less than ten dollars for individual single-station battery-operated smoke alarms to thousands of dollars for complete addressable analog systems. Individual detectors typically cost between $100 and $150 in commercial systems and control panels typically cost up to a few thousand dollars, depending on the size of the system. Including installation and control panel costs, a total cost of approximately $300 per detector is sometimes used for estimating purposes.

Additional information on current fire detection and alarm technologies is available at the websites of the manufacturers. Websites for some manufacturers of fire detection and alarm equipment are identified below:

http://www.cerbpyro.com/
http://www.firstalert.com/
http://www.notifier.com/nfs_home.htm
http://www.simplexnet.com/products/fire/
http://www.worldelectronics.com/

In summary, fire detection and alarm systems are an essential element of a fire protection program for a residential building. These systems provide early notification of fire development. In cases of relatively slow fire development, these systems usually provide sufficient warning to permit effective intervention or evacuation. Regardless of the type of fire alarm system that is installed, however, these systems do nothing by themselves to alter the development of a fire. Consequently, in cases of relatively fast fire development and in cases where people may not hear or react to an alarm signal, fire alarm systems have limited value. Unfortunately, many fires develop too fast for effective suppression by the fire department before they become hazardous, even with prompt detection and notification. In these cases, automatic fire suppression is desirable.

Suppression

The final element of the PODS concept is suppression. The best weapon for controlling a fire in its early stages, before it becomes too hazardous, is the automatic sprinkler system. An automatic sprinkler system is an integrated system of underground and overhead piping connecting one or more automatic water supplies, such as a city water main, with automatic sprinklers. While a sprinkler system may appear to be a fairly ordinary plumbing system, it is not. An automatic sprinkler system is a specialized system requiring professional skills for design, installation and maintenance in conformance with recognized standards.

The distinguishing feature of an automatic sprinkler system is the automatic sprinkler itself. Automatic sprinklers are closed nozzles that hold back water under pressure within the pipes, much like a faucet does when it is closed. Special sprinklers have been designed for residential occupancies. These devices are more sensitive than traditional sprinklers used in commercial and industrial buildings; they are designed to respond before life-threatening conditions develop within the room where the fire starts and thus provide protection even for people sleeping in that room.

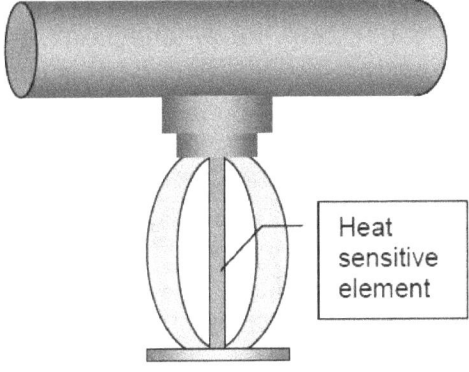

Heat sensitive element

Contrary to popular portrayals in the media, sprinkler systems do not activate when somebody pulls the fire alarm or when a smoke detector activates. Nor do all sprinklers open simultaneously when the first one activates. Each sprinkler has its own heat sensitive element that must be heated to its activation temperature of about 165°F before the sprinkler will operate. This heat sensitive element holds a cap in place over the sprinkler's orifice.

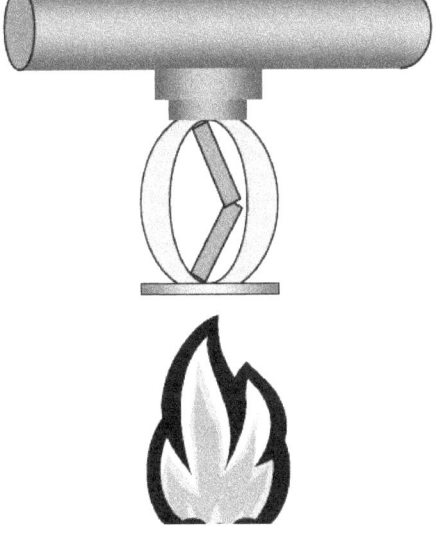

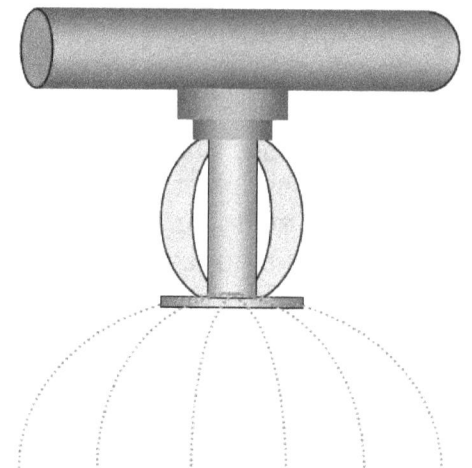

Once activated, a sprinkler discharges water at a rate of approximately 20 gallons per minute in a fairly uniform spray pattern throughout the room. This water spray cools the fire environment while wetting fuels surrounding the fire source to prevent or retard their ignition. In this way, the fire is held in check until the fire department can respond and complete extinguishment of the fire.

Typical dormitory and fraternity sleeping rooms only require the installation of a single sprinkler for protection. Special sidewall sprinklers can be installed near the entrance to each room, just off the corridor. This simplifies the installation process, while minimizing the aesthetic impact. The only part of the system evident in a room is the sprinkler itself.

With this arrangement, supply pipes can be installed at the ceiling along the length of each corridor to serve every room. These pipes also serve sprinklers in the corridor. The supply pipes in the corridor can then be enclosed to minimize the aesthetic impact of the installation. Special covers have been designed and developed for this purpose. These covers simply snap in place to cover the pipes while providing convenient access when needed.

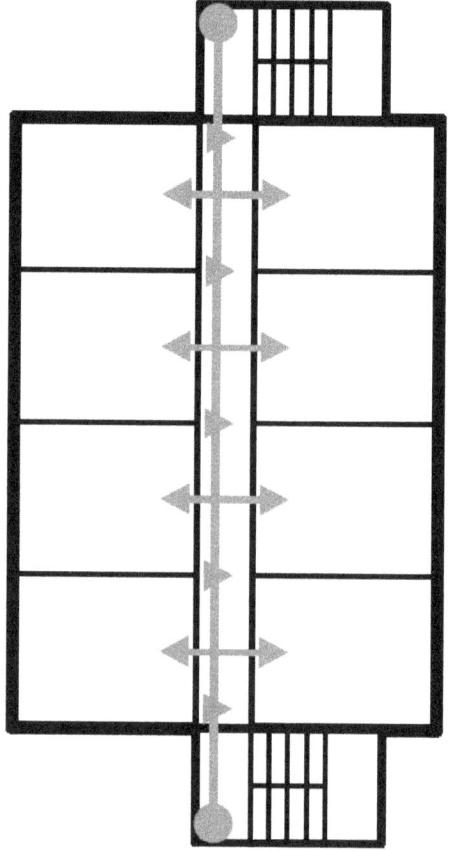

Sprinkler pipe cover

Delivery of water to each corridor is via "riser" pipes. These risers are typically installed, along with drainage pipes and system test assemblies, in exit stairways, where aesthetics are not as important. In many buildings, standpipes are already located in the stairways and can be converted to combined sprinkler-standpipe risers.

Large common areas in student housing, including living and dining rooms, may require the installation of more than one sprinkler for proper coverage. Regardless of the number of sprinklers installed in a space, however, the vast majority of residential fires are controlled by the operation of only one or two sprinklers; less than 10 percent of residential fires require the operation of more than four sprinklers. Thus, the vast majority of fires in sprinklered residential buildings will be controlled with a water flow rate of less than 50 gallons per minute.

Compared with the 150 to 250 gallons per minute discharged by a single fire department hose stream, concerns about water damage from sprinkler operation during a fire are put into proper perspective. Sprinklers respond much earlier than the fire department possibly can and they concentrate water delivery where it is needed. Fires remain much smaller with sprinkler protection, such that the combined effects of fire and water damage are generally much lower than in unsprinklered buildings. Recovery after a fire is typically much quicker as well.

Sprinkler leakage is a potential cause of water damage that raises concern among facilities managers. Accidental leakage from sprinklers or piping is extremely rare, about the same as for the other plumbing systems in a building, but sprinklers can be mechanically damaged to cause leakage. Unfortunately, student horseplay is a fairly common cause of such damage in student housing facilities. At the University of Maryland, for instance, campus safety authorities have come to fear the 3 F's - Freshmen, Footballs and Frisbees - because of the occasional sprinkler that has been struck and activated by students playing "catch" in the halls. Students sometimes use sprinklers to hang clothing or decorations as well, activities that must be discouraged. Certain features can be designed into a sprinkler system to minimize the potential for such damage, including the use of concealed sprinklers, but students should be made aware of the consequences of their actions. Some universities hold students financially accountable for damages caused by inappropriate activities.

Ultimately, the decision to install automatic sprinkler systems in student housing should be based on a number of factors, including the costs and benefits associated with the installation of such systems. However, not all these factors are tangible, while others are difficult to quantify, so the decision to install sprinklers may not be based purely on an economic analysis.

In general, the costs and inconveniences associated with the installation of sprinkler systems are lower in new construction than in retrofit applications. Current nationally recognized fire safety codes and standards generally require the installation of automatic sprinkler systems in new dormitories and apartments, so such installations would be prudent even if not yet mandated by local regulations. This discussion will focus on retrofit applications.

Direct costs associated with automatic sprinkler systems include initial design and installation costs as well as ongoing inspection and maintenance costs. The design and installation of sprinkler systems in retrofit applications generally costs in the range of $2.00 to $3.00 per square foot, depending on locale and conditions. For comparison, this is similar to the cost of installing wall-to-wall carpeting in a building. These costs can be reduced if the installation is coordinated with a major building renovation.

Sprinkler systems are fairly rugged mechanical systems that do not require a lot of maintenance, but it is important that they be inspected regularly in accordance with recognized standards to assure they are ready to operate in the event of a fire. Sprinkler system inspection and maintenance typically costs between about $250 per year and $1,000 per year, depending on the size of the building. Smaller facilities, such as fraternity houses, would be near the low end of the scale, while large facilities, such as high-rise dormitories, would be near the high end. Annual testing of fire pumps typically costs about $500 per pump.

Some buildings, particularly high-rise dormitories, may require the installation of a fire pump to boost the pressure of the incoming water supply because water pressure decreases inversely with building height. Space for the pump and its associated equipment will need to be provided, typically in a basement or utility space. Fire pumps contribute significantly to the cost of a sprinkler system, with installed costs in the neighborhood of $25,000 fairly common for this equipment.

Indirect costs associated with the installation of sprinkler systems include items such as water connection fees and property taxes. Different jurisdictions have different requirements regarding these fees, so it is difficult to generalize about them. Some jurisdictions do not impose these fees for automatic sprinkler protection as one way to encourage the use of sprinklers in buildings.

Direct benefits associated with the installation of automatic sprinkler systems in student housing include lower risks of life loss and property damage from fire. These benefits may translate into reduced insurance premiums as well, helping to defray the costs of sprinkler protection. In some cases, these reduced premiums translate into payback periods of 5- to 10-years, so that the sprinkler protection eventually pays for itself.

Less tangible but very real benefits include the good will and peace of mind associated with providing students with this high level of fire safety. When a fire inevitably does occur, clean up is also much easier so down time and student displacement are minimized after a fire. This can be an important factor in the middle of the school year.

Once the decision is made to install sprinklers in your student housing facilities, it is necessary to procure professional design and installation services to complete the job. Only properly qualified and certified fire safety professionals should be entrusted with this job. Competition in the sprinkler industry tends to be strong, so it will usually be in your best interest to obtain multiple bids for a job. But your selection of a contractor should not be based on cost alone; references, reputations, credentials and insurance should also be considered to assure that a contractor is capable and qualified to complete your job.

The contractor will obtain appropriate permits and approvals from the authority having jurisdiction before proceeding with the installation. Your insurance representative should also be involved in the review process to assure maximum credit on your fire insurance premium.

The primary inconvenience associated with retrofitting sprinklers into an occupied building is the installation process itself. Fortunately, the hospitality industry has already addressed this issue and has developed installation materials and techniques that can minimize this impact. Scheduling the installation during a major building renovation or during a period when the building is not occupied, such as during winter or summer breaks, can further minimize this impact, but even when a building is occupied, the installation can proceed smoothly. In the hospitality industry, it is not unusual for systems to be installed while the building is occupied. With the installation techniques described above, most of the work is in the stairways, corridors and other public areas of a building; the installation time required in individual rooms is only about an hour or two.

When the installation is completed, the contractor must test the system in accordance with recognized standards. This testing should be conducted after notification of the authority having jurisdiction and the owner's representative so these parties can be present to witness the testing if they desire. Once the authority having jurisdiction and the owner have accepted the installation, responsibility for maintaining the system transfers to the owner or operator of the building. Proper maintenance requires periodic inspection and testing of system components in accordance with recognized standards.

Qualified employees or independent agencies can perform these activities. The contractor who installed the system may provide such services under an annual service contract. Otherwise, many fire safety professionals are qualified to perform these essential services.

Summary

There are many issues vying for the attention of, and limited resources available to, campus housing authorities and fraternal organizations. Rational decisions need to be made with respect to allocating these limited resources to best serve the students and the campus community. Keys to the decision-making process include a clear understanding of the risk factors involved, the different risk reduction methods and strategies available, and the costs and benefits of the different options. The purpose of this report has been to introduce an integrated approach to fire safe student housing comprised of Prevention, Occupant awareness and training, Detection and alarm, and Suppression. You can remember these elements by the acronym "PODS."

Serious fires in student housing wreak almost unimaginable devastation and disruption; this potential warrants careful consideration of fire safety options. In particular, automatic sprinkler systems should be considered as a viable option; they have established an impressive record of preventing residential fire catastrophes, particularly in the hospitality industry. With technologies developed specifically for residential applications, automatic sprinkler systems are now commonplace in hotels and motels, where they afford the traveling public with a high level of fire safety. College students deserve this same high level of protection ... and their parents expect it.

In closing, it is reiterated that fire safe student housing does not just happen; it requires careful planning, coordination, implementation and diligence on the part of campus housing and safety administrators.

Acknowledgements

This work was sponsored by the United States Fire Administration of the Federal Emergency Management Agency under Contract No. EME-97-CA-0333. Thanks to Mr. Larry Maruskin, the Project Manager for the USFA, for his guidance on this project. Special thanks to Mr. Phil Friday, a graduate student in the Department of Fire Protection Engineering at the University of Maryland, for stepping in and looking after the details to bring this project to completion. The videotape that accompanies this report was developed under the direction of Ms. Sonya Sezun and Mr. Mac Nelson of the Video Production unit of the University Marketing department at the University of Maryland. Some of the still images in this report were obtained from the website maintained by the Building and Fire Research Laboratory of the National Institute of Standards and Technology (www.bfrl.nist.gov).

Appendix A

AN EXAMINATION AND ANALYSIS OF SELECTED COLLEGE DORMITORY AND FRATERNITY HOUSE FIRES

By

Dr. John L. Bryan
Professor Emeritus
Department of Fire Protection Engineering
University of Maryland

December 5, 1997

A1. INTRODUCTION

Nine college dormitory and six fraternity house fires for a total of fifteen fire incidents that occurred from April 5, 1967 until May 12, 1996 were selected from the review of literature for specific analysis and study. These fraternity and college dormitory fire incidents were selected based on the criteria of the availability of a published report and the occurrence of fatalities or injuries to the occupants. Thus, college dormitory and fraternity house fires illustrated in published reports without occupant fatalities or injuries are not included in this examination.

These fifteen fire incidents involved a range of facilities relative to their locations varying from large metropolitan areas to rural areas and thus the type of public fire protection available. Four fire incidents occurred in facilities located in large metropolitan areas with career fire departments including: Washington, D.C. (3), Providence, RI (6) and Columbus, Ohio (12) (19). Five fire incidents occurred in facilities located in suburban or small city areas with career or combination fire departments including: Amherst, MA (1), Baltimore County, MD (4), Berkeley, CA (9), Chapel Hill, NC (10)(22), and Cambridge, MA (11).

Six fire incidents occurred in facilities located in small town or rural areas with combination or volunteer fire departments including: Dover, DE (5),

Farmville, VA (5), Cayuga Heights, NY (7), Saratoga Springs, NY (13), Baldwin

City, KS (14), and Bloomsburg, PA (21).

This examination and analysis will review in the first section the factors

identified in the investigations to have been significant in the ignition of the fire

and the propagation of the fire resulting in the fatalities or injuries to the occupants.

The second section of this report will examine the construction, age and height

variables of the facilities with the number and arrangement of the exit stairways.

The third section of this analysis will examine the characteristics of the occupant

population and the human behavior variables of the occupants identified as

contributing to the fire ignition, the fire development, the fatalities and injuries and

jumping behavior. The fourth section of this report will review the information

from the investigation reports relative to the arrangement of the fire protection

systems in the facilities. The fifth section of the report will review additional

NFPA data on college dormitory and fraternity house fires. The final section of

this report will provide a summary of the physical, psycholgical and cultural

factors with the recurring variables that appear to be significant in the college

dormitory and fraternity house fires. This final section will also include an

examination of the similarities and the differences in the fire incidents in these two

unique and specific occupancies housing college age occupants.

A2. IGNITION AND PROPAGATION VARIABLES

The nine dormitory fire incidents and the six fraternity house fires are identified in Table 1 by the name of the educational institution operating the dormitory facility or from which the students were members of the fraternity. Nine of these educational institutions were private facilities: Amherst College (1), Baker University (14), Cornell University (7), George Washington University (3), Longwood College (5), Massachusetts Institute of Technology (11), Providence College (6), Skidmore College (13), and Wesley College (5). In addition to these nine private institutions there were five state supported institutions involved in the remaining six fires with Ohio State University having two fires in this study population: Bloomsburg State College (21), Ohio State University (12)(19), University of Maryland Baltimore County (4), University of California Berkeley (9) and The University of North Carolina (10).

The six fraternity house fires involved students from Amherst College, Baker University, Bloomsburg State College, Ohio State University, University of California Berkeley and the University of North Carolina. Thus, the fraternity house fires involved students from two private institutions and four state supported institutions. The nine dormitory fires occurred on the campuses of Cornell University, George Washington University, Longwood College, Massachusetts Institute of Technology, Ohio State University, Providence College, Skidmore

College, University of Maryland Baltimore County and Wesley College. These dormitory fires occurred on the campuses of seven private institutions and two state supported institutions.

It is also of interest as indicated in Table 1, that of these fifteen fire incidents thirteen, or 86 percent of the fires occurred in the early morning hours between 1:00 am and 7:00 am. Only two of these fires both in dormitories occurred outside this time frame, one at 11:24 am and the other at 10:59 p.m. (11)(19)

Relative to the ignition factors the predominate factors appeared to be incendiary and smoking materials. Five of these fire incidents were attributed to incendiary actions, four involving college dormitories (3)(4)(5)(19) and one involving a fraternity house. (9) Four of these fire incidents were identified as involving smoking materials, all of these identified smoking caused fires involved fraternity houses. (1)(10)(14)(21) In four of these fire incidents the investigators could not determined the ignition source. The remaining two fire incidents were determined to be caused by burning decorations in a fireplace (12) and an overloaded extension cord. (5)

Upholstered sofas were the most predominate fuel immediately ignited in seven of the fire incidents, contributing to the rapid fire propagation in fires involving five fraternity houses (1)(9)(12)(14)(21) and two college dormitories. (7)(19) The second most readily involved fuel material consisted of trash being ignited in three of the fire incidents, involving two college dormitories (4)(11) and one fraternity house. (10) Wood paneling was identified as one of the principal factors contributing to the fire propagation in five of the fire incidents involving four fraternity houses (1)(9)(10)(14) and one college dormitory. (7) Paper decorations and wall coverings were identified as increasing the fire propagation in four fire incidents involving three college dormitories (6)(11)(13) and one fraternity house. (12) Fire propagation was accelerated by carpeting in two college dormitory fire incidents. (3)(4)

A construction aspect presented in Table 2 relative to the fire propagation of flame and smoke spread to the upper floors was enhanced by open stairways or enclosed stairs with the doors blocked opened or removed in seven of these fire incidents. These seven fire incidents involved all six fraternity houses and one college dormitory. (7) In addition, one fraternity house fire spread was aided by the operation of a large fan in the basement providing a mechanical draft up the single open stairway. The construction variables of the fifteen dormitory and fraternity house structures involved in the study population will be examined and analyzed next involving the stairways and stair enclosures.

Table 1		
Facility Date & Time	**Ignition Factors**	**Propagation Variables**
Cornell University (7) Dorm. 4-5-67 0400	Undetermined, source in basement lounge area	Plywood paneling & sofas
Ohio State University (19) Dorm. 5-22-68 2259	Incendiary in living room of suite 1140	Sofa, spread into room 1142 through open door
Amherst College (1) Frat. 2-2-75 0559	Smoking & wastebasket to sofa in living room	Sofa, oak paneling & open 1st. floor & stairs
Mass. Inst. of Tech. (11) Dorm. 7-22-75 1124	Undetermined, trash in hall by trash chute	Trash in hall & vinyl wall covering
Ohio state University (12) Frat. 1-8-76 0202	Burning decorations in fire-place ignited trash & sofas	Paper decorations, trash & sofas
Skidmore College (13) Dorm. 4-5-76 0400	Undetermined, in trash closet on 1st. floor	Vinyl Wall covering
Baker University (14) Frat. 8-29-76 0306	Smoking on a sofa in TV room	Wood paneling from TV room to open stair
Providence College (6) Dorm. 12-13-77 0257	Undetermined, in room 406	Christmas decorations on walls & doors
George Washington U. (3) Dorm. 4-19-79 0345	Incendiary, flammable liquid on hall carpet	Carpet & flammable liquid in hall on 5th. floor
U. of MD. Balt. County (4) Dorm. 2-3-80 0359	Incendiary, trash in hall & trash room, 2nd. floor	Trash & carpet in 2nd. floor hall
Wesley College (5) Dorm. 4-12-87 0233	Incendiary, smoke bomb in room 206	Room furnishings
Longwood College (5) Dorm. 4-28-87 0650	Overloaded extension cord in 3rd. floor room	Hanging bed linens
U. Cal. Berkeley (9) Frat. 9-8-90 0100	Incendiary, lighter on sofa in living room	Wood paneling throughout including stairs
Bloomsburg State C. (21) Frat. 10-21-94 0502	Smoking on sofa	Sofa believed out & placed on porch & open stair
U. North Carolina (10) Frat. 5-12-96 0607	Smoking material & trash Under bsmt. bar	Wood paneling in bsmt. open doors & fan in bsmt.

A3. CONSTRUCTION VARIABLES

The construction variables relative to the nine college dormitory and six fraternity house buildings are presented in Table 2, consisting of the type of construction, the age and height of the building, the number and enclosure of the stairways. The age of the buildings at the time of the fire was not reported for three fraternity houses and for one college dormitory. The range of age for the eleven reporting facilities involved in these fire incidents varied from 70 years to 9 months, with a mean age of 29.7 years. The range of age for the seven college dormitories varied from 39 years to 9 months with a mean age of 15.5 years while the age range for the four reporting fraternity houses varied from 33 to 70 years with a mean age of 54.7 years. Thus, the fraternity houses when comparing mean ages were more than three times as old as the college dormitories.

Relative to the height of the structures as would be expected the college dormitories were the taller structures varying in height from 2 stories and basement to 24 stories, with a median height for the nine college dormitories of 9.1 stories. The six fraternity houses varied in height from 2 stories to 3 stories and basement, with a mean height of 2.5 stories as would be expected, since many of the fraternity houses were converted single or multi-family residential structures.

There was also a definite difference in the construction of the college dormitory buildings compared to the fraternity houses. Eight of the college dormitories were of fire resistive construction and one was of protected noncombustible construction. (6) Two of the fraternity houses were of frame construction, (9)(21) three were of ordinary construction (10)(12)(14) and one was of protected noncombustible construction. (1) All of the fraternity houses were located off campus and all of the college dormitories were located on campus.

Table 2			
Facility	**Age-Years**	**Exit Stairways**	**Height-Construction**
Cornell University Dorm. (7)	14	2 enclosed stairs, doors removed & open	2 stories & bsmt., fire resistive
Ohio State U. Dorm. (19)	0.75	2 enclosed stairs	24 stories, fire resistive
Amherst College Frat. (1)	57	3 open stairs	2 stories & bsmt., prot. Noncombustible
Mass. Inst. of Tech. Dorm. (11)	Not reported	2 enclosed stairs	24 stories, fire resistive
Ohio State U. Frat. (12)	Not reported	1 open stair, fire escape- 2nd. floor	2 stories & attic, ordinary
Skidmore College Dorm. (13)	10	2 enclosed stairs, 1 enclosed conv. stair	3 stories & bsmt., fire resistive
Baker University Frat. (14)	59	1 enclosed stair-open doors & 1 open stair	2 & 3 stories & bsmt., ordinary
Providence College Dorm. (6)	39	3 enclosed stairs	4 stories, Prot. noncombustible
George Wash. U. Dorm. (3)	Not reported	2 enclosed stairs	9 stories & bsmt., fire resistive
U.of MD. Balt. Cty. Dorm. (4)	10	3 enclosed stairs	3 stories & bsmt., fire resistive
Wesley College Dorm. (5)	18	2 enclosed stairs	3 stories & bsmt., fire resistive
Longwood College Dorm. (5)	17	Not reported	10 stories & bsmt., fire resistive
U. Cal. Berkeley Frat. (9)	33	3 enclosed stairs-open doors	3 stories & bsmt., wood frame
Bloomsburg St. C. Frat. (21)	Not reported	1 open stair,	2 stories, wood frame
U. North Carolina Frat. (10)	70	1 open stair, 2 fire escapes-2nd. & 3rd. fls.	3 stories & bsmt., ordinary

A4. OCCUPANT BEHAVIOR VARIABLES

These nine college dormitory and six fraternity house fires were selected

from the criteria that each fire resulted in occupant fatalities or injuries. Six of the

college dormitory fires resulted in fatalities in a range from 1 to 10, with a total of

24 fatalities. Five of the fraternity house fires resulted in fatalities in a range from

2 to 5, with a total of 20 fatalities. The injuries to the occupants in the nine college dormitory fires varied in a range from 0 to 60, with a total of 133 occupants suffering medically treated injuries. One college dormitory fire resulted in a single fatality and no injuries. The injuries to the occupants in the six fraternity house fires varied from 0 to 4, with a total of 10 occupants suffering medically treated injuries. In two of the fraternity house fires, each resulting in 5 occupant fatalities, there were no medically treated occupant injuries.

Occupants jumping from the windows has been identified as a phenomenon of the occupant behavior during college dormitory and fraternity house fires. Two of the occupant fatalities resulted from jumping and six of the occupant injuries were identified as resulting from jumping behavior. A total of 17 occupants were identified as having engaged in jumping behavior with jumps varying from the second, third, fourth and fifth floors.(3)(4)(6)(7)(9)(10)(12)(14) Additional occupants were reported to have jumped from windows in one fraternity house fire and one dormitory fire. (1)(13) Jumping behavior was reported to have occurred in five or 55 percent of the college dormitory fires and in five or 83 percent of the fraternity house fires. Occupants were rescued from windows with fire department ladders in three of the college dormitory fires, (3)(5)(6) and in two of the fraternity house fires. (1)(9)

The total number of occupants in the facility at the time of the fire was often difficult for the investigators to obtain due to the transient nature of the occupants in both the college dormitories and the fraternity houses. However, the reported number of occupants was generally larger in the dormitories due to the larger size of the facilities. The number of occupants for the seven college dormitories reporting varied from 84 in a 3 story dormitory to 1,950 occupants in a 24 story dormitory. The number of occupants in the four reporting fraternity houses at the time of the fire varied from 9 to 39 occupants. The gender of the occupants in the six college dormitories reporting this data were mixed in four of the six dormitories with two dormitories having all female occupants, resulting in all female fatalities. (6)(13) All six of the fraternity house fires reported the gender of the occupants at the time of the fire. In these six fraternity houses there were all male occupants in two of the houses at the time of the fire. (1)(14) Four of the fraternity houses contained male and female occupants with both male and female fatalities. (9)(10)(12)(21)

Occupant behavior involved with the alerting of the other occupants of the facility was reported in 14 of the 15 fire incidents. This behavior consisted of shouting, knocking on doors and activation of the manual fire alarm stations. Occupant fire fighting behavior occurred in only two fraternity house fire incidents. In one case a occupant attempted to utilize a fire extinguisher unsuccessfully and in the other case a couch believed to have been extinguished was moved to an outside porch, where it later reignited. (12)(21) This occupant behavior is summarized in Table 3.

Table 3					
Facility	**No.**	**Gender**	**Fat.**	**Inj.**	**Jumping**
Cornell University Dorm. (7)	69	No report	9	2	Several from 2nd. floor
Ohio State U. Dorm. (19)	1,940	Mixed	2	1	None
Amherst College Frat. (1)	39	Male	0	4	6 from 2nd. floor
Mass. Inst. of Tech. Dorm. (11)	No report	Mixed	1	0	None
Ohio State U. Frat. (12)	22	Mixed	2	1	2 from 2nd. floor
Skidmore College Dorm. (13)	84	Female	1	60	Numerous from all 3 floors
Baker University Frat. (14)	No report	Male	5	0	1 from 3rd. floor
Providence College Dorm. (6)	41-4th fl. Est. 160	Female	10	12	2 from 4th. floor
George Wash. U. Dorm. (3)	898	Mixed	0	37	2 from 5th. floor
U.of MD. Balt. Cty. Dorm. (4)	312	Mixed	0	2	1 from 2nd. floor
Wesley College Dorm. (5)	180	No report	1	4	None
Longwood College Dorm. (5)	No report	No report	0	15	None
U. Cal. Berkeley Frat. (9)	10	Mixed	3	2	1 from 2nd. floor
Bloomsburg State C. Frat. (21)	9	Mixed	5	0	None
U. North Carolina Frat. (10)	No report	Mixed	5	3	2 from 3rd. floor
Total	3,723		44	143	17+

A5. PROVISION OF FIRE PROTECTION SYSTEMS

None of the nine college dormitories or the six fraternity houses were completely equipped with an automatic sprinkler system. Two of the college dormitories that were high-rise structures were partially sprinklered. One of these dormitories had sprinklers provided in the storage and service areas. (11) The other dormitory had a complete suite arrangement, with the suite's living rooms, kitchens and corridors sprinklered, and the student's sleeping rooms non sprinklered. (19) Two of the fraternity houses had partial sprinkler systems. One of the fraternity houses had sprinklers throughout the basement of the facility. (1) The other fraternity house had a single sprinkler at the top of the trash chute and at the top of the laundry chute. (9)

Eight of the college dormitories had a local fire alarm system with manual stations and one college dormitory had no fire alarm system. (7) The fire alarm system at one of these dormitories was connected directly to the fire department (6) and two were connected directly to campus communication centers.(11)(19) Of the eight fires involving college dormitories with alarm systems, the systems were activated in six of the fires. The system failed to operate in one dormitory fire and the system was turned off in another dormitory at the time of the fire. (5)

In one college dormitory fire the manual station was activated by the fire exposure (11) and in another college dormitory fire the system was deactivated by the fire exposure following the initial activation. (3)

Considering the six fraternity houses in this study population, only two of the houses had fire alarm systems. The manual fire alarm system was not activated in one of the fires (9) and in the other fire the thermal detectors connected to the system activated the system. (1) The fire alarm system in this fraternity house had rate-of-rise and fixed-temperature detectors installed throughout the house and the fire alarm system activation was transmitted directly to the fire department. (1)

Five of the college dormitories were reported as having either smoke or thermal detectors installed in portions of the buildings. Two of the college dormitories had smoke detectors in each students room. (4)(5) In the fire in one of these dormitories the smoke detector failed to operate in the room of fire origin, and subsequent investigation determined that 85 percent of these detectors in the dormitory were inoperative. (5) One of the college dormitories had thermal detectors of the combination rate-of-rise fixed-temperature type in each students room and the basement trash room, the fire started in the first floor trash room. (13) One college dormitory had thermal detectors of the rate-of-rise fixed-temperature type installed at the top of each stairway. (6) One college dormitory had smoke detectors in all the trash rooms and at the top of the trash chutes and exhaust ducts.

(11) Considering the six fraternity houses involved with fires in this study, two of the houses had no detectors as well as no alarm system.(12)(14) Two of the fraternity houses had partial installations of smoke detectors, one with single-station detectors in some sleeping rooms. (9) The other fraternity house had thermal detectors in the basement mechanical room and smoke detectors in the basement stairway and the second and third floor corridors. (10) One fraternity house was reported to have had both thermal and smoke detectors which apparently failed to operate at the time of the fire incident. (21) One fraternity house was completely protected with thermal detectors of the rate-of-rise fixed-temperature type and smoke detectors at the top of the stairs connected directly to the fire department which activated in the fire incident. (1) The facilities fire protection systems are summarized in Table 4.

Table 4			
Facility	**Alarm System**	**Detectors**	**Sprink.- Standp.**
Cornell University Dorm. (7)	None	None	None
Ohio State U. Dorm. (19)	Manual, sprink. act. to campus PD	None	Standp.- stairs, stor. & ser. areas sprink.
Amherst College Frat. (1)	Manual, detectors & sprink. act. to FD	Smoke-stairs, thermal throughout	Basement-sprink.
Mass. Inst. of Tech. Dorm. (11)	Manual, detectors & sprink. act to Con.C	Smoke- top of exh., trash chutes &rooms	Suite ex.stud. rms. trs.chs. standp-strs.
Ohio State U. Frat. (12)	None	None	None
Skidmore College Dorm. (13)	Manual, detectors act. bldg. alarm	Thermal-bsmt. trash rm. & student rooms	None
Baker University Frat. (14)	None	None	None
Providence College Dorm. (6)	Manual, detectors act. to FD	Thermal top of stairs	None
George Wash. U. Dorm. (3)	Manual	No report	Standp. - stairs
U. of MD.Balt. Cty. Dorm. (4)	Manual	Smoke - student rooms	Standp. - stairs
Wesley College Dorm. (5)	Manual	None	None
Longwood College Dorm. (5)	Manual	Smoke - student rooms	Standp. - stairs & hose cabinets
U. Cal. Berkeley Frat. (9)	Manual	Smoke-some student rooms	Standp. - hose cabinets
Bloomsburg State C. Frat. (21)	None	Smoke & thermal detectors	None
U. North Carolina Frat. (10)	None	Thermal-bsmt.rm. smoke-stair & hall	None

A6. NFPA DATA ANALYSIS

The National Fire Protection Association has published data identifying the established causes of the fires from a review of 260 college dormitory, fraternity and Sorority house fires in 1954. (17) In 1995 the NFPA published data relative to the established causes of the fires in college dormitories separately from the fraternity and sorority house fires for the fires reported from 1990 through 1994. (16) It should be recognized the causes of the fires published in 1954 were based on the ten year time period from 1944 through 1954.

Although there are obviously significant social and cultural differences in both the institutions and the students, a comparison of the fire cause data between this approximate 40 year interval appeared to be at least of historical value. In the 1940's and 1950's there were more restrictions on the student's behavior in both college dormitories and the fraternity and sorority houses with hour limitations and adult supervision consisting of "house directors" being the norm. It should also be remembered in both 1944 and 1945 a large percentage of the college age population were in the military service, and in the late 40's and early 50's this population was in college providing a culturally different population from the college students of the early 90's. There would also appear to be some construction differences with the use of high-rise dormitories, since the 1970's with enclosed stairways and fire resistive construction. Obviously, smoke detectors were not available until the early 1960's for residential occupancies.

Table 5 presents the causes of the college dormitory, fraternity and sorority house

fires as published by the National Fire Protection Association in 1954. (17)

Table 5	
NFPA 1954 Data	
Causes of Fires	Percentage of Fires
Smoking and matches	24.2
Misuse of electricity	22.7
Defective chimney	10.0
Heating equipment, defective	9.6
Incendiary or suspicious	9.6
Spontaneous ignition	6.2
Kitchen hazards	5.8
Open fireplaces	5.3
Lightning	0.8
Exposure	0.8
Explosion	0.8
Miscellaneous	4.2
Total	100.0

The National Fire Protection Association has presented the data separately

for college dormitory, fraternity and sorority house fires for the period from 1990

to 1994. This data is also developed in the form of an annual average for the five

year period estimated from both NFPA and NFIRS (National Fire Incident

Reporting System) data. It should be remembered the NFIRS data is generated by

the incident reports of the fire department agencies involved in each fire incident.

Thus, fire incidents not reported to a fire department would be excluded. Table 6

presents the causes of fire data with the dollar loss as developed and published by

the NFPA for fraternity and sorority houses in 1995. (16)

NFPA Causes of Fires	Table 1995 No.	6 Data %	Dollar Loss	%
Incendiary, suspicious	33.89	19.5	497,497.00	20.1
Cooking	25.93	14.9	15,759.00	0.6
Electrical distribution	16.90	9.7	822,738.00	33.3
Smoking	15.62	9.0	126,790.00	5.1
Other Equipment	12.04	6.9	12,729.00	0.5
Other heat, flame, spark	10.25	5.9	156,178.00	6.3
Open flame, ember, torch	9.35	5.4	19,463.00	0.8
Heating	8.96	5.2	296,958.00	12.0
Appliances, air conditioning	8.46	4.9	54,364.00	2.2
Natural causes	1.78	1.0	179.00	0.0
Children playing	0.90	0.5	354.00	0.0
Exposure	0.89	0.5	0.00	0.0
Unknown	28.53	16.4	466,528.00	18.9
Total	173.49	100.0	2,469,538.00	100.0

Table 7 presents the identical fire cause categories as Table 6 with the NFPA

1995 fire cause and dollar loss data for college and University dormitories. (16)

NFPA Causes of Fires	Table 1995 No.	7 Data %	Dollar Loss	%
Incendiary, suspicious	428.92	29.9	2,231,154.00	40.0
Cooking	225.06	15.7	116,815.00	2.1
Smoking	155.54	10.8	188,031.00	3.4
Other equipment	86.42	6.0	262,063.00	4.7
Appliances, air conditioning	85.13	5.9	663,876.00	11.9
Electrical distribution	77.16	5.4	520,370.00	9.3
Open flame, ember, torch	73.03	5.1	78,778.00	1.4
Other heat, flame, spark	70.48	4.9	210,223.00	3.8
Heating	35.69	2.5	251,368.00	4.5
Exposure	8.52	0.6	3,887.00	0.1
Natural causes	7.16	0.5	63,422.00	1.1
Children playing	7.13	0.5	3,834.00	0.1
Unknown	176.30	12.3	980,059.00	17.6
Total	1,436.54	100.0	5,573,879.00	100.0

Comparing the 1995 NFPA data for fraternity and sorority houses to the

identical data for college and university dormitories identifies some interesting

differences and similarities. As can be noted, the greatest causes of fires in both

occupancies are "incendiary, suspicious" and "cooking" accounting for 34.4

percent of the fires in fraternity and sorority houses and 45.6 percent of the fires in the college dormitories. "Smoking" is the fourth leading cause of fire in the fraternity and sorority houses following "electrical distribution" as the third leading cause. When these four leading causes of fire are considered they account for 53.1 percent of all the fires during this 1990 through 1994 period as reported by the NFPA and thus, should be the focus of the fire prevention efforts in these occupancies.

In the college dormitories "smoking" is the third leading cause of fires followed by "other equipment" as the fourth leading cause of fires. Thus, in the college dormitories these four causes of fire account for 62.4 percent of all the fires. However, since the "other equipment" cause accounts for only 6 percent of the fire causes, the remaining three leading causes of fire consisting of "incendiary, suspicious", "cooking" and "smoking" account for 56.4 percent of all the fires in college dormitories. Obviously, for both the fraternities and sorority houses and the college dormitories the principal causes of fire are the human behavior related categories of "incendiary, suspicious," "cooking" and "smoking" and these are the causes of fire that should be alleviated by the fire prevention related efforts of both education and enforcement. The final two sections of this report will contain specific suggestions for the development of programs and efforts in these areas.

When one compares the NFPA data on fire causes reported in 1954 as illustrated in Table 5 with the leading causes from the 1995 data presented in Tables 6 and 7 there appear to be some definite differences. It must be remembered of course that the 1954 data is a compilation of both college dormitory fires and fraternity and sorority house fires so the compilation and presentation of the data is basically different from the 1995 data. However, there are some interesting similarities and differences in the fire causes as reported with this forty-year difference.

The category of "incendiary or suspicious" in the 1954 data was the fifth leading cause of fires accounting for only 9.6 percent of all the fires. While in the 1995 data for the fraternity and sorority houses and the college dormitories this was the leading cause of all the fires accounting for 29.9 percent of the fires in the college dormitories and 19.5 percent of all the fires in the fraternity and sorority houses. This is an increase of at least 100 percent in the fraternity and sorority houses and approximately 200 percent in the college dormitories during this forty-year period. In the 1954 data the category of "misuse of electricity" was the second leading cause of fires accounting for 22.7 percent of all the fires. While in the 1995 data the category of "electrical distribution" was the third leading cause of fires in the fraternity and sorority houses it only accounted for 9.7 percent of all the fires in these occupancies. The 1995 data for college dormitories indicated the category of "electrical distribution" was the sixth leading cause of fire accounting

for only 5.4 percent of all the fires in college dormitories. It appears the higher

ranking of electrical fires in the 1954 data may be a result of two variables. First,

as previously indicated the 1954 data included the fraternity and sorority house

data with the college dormitory data, and as the 1995 data confirms electrical

caused fires are still the third leading cause of fires in fraternity and sorority

houses. It may be there is still more utilization of unapproved electrical

modifications and extensions in the fraternity and sorority houses than in the

college dormitories which in general appear to have newer structures and more

supervision. Secondly, it should be recognized the determination of fire cause has

become more technical and sophisticated during the forty-year period from 1954 to

1995, and with the development of the NFIRS reporting system the overall validity

of the identification of the fire causes has probably improved.

The 1954 data indicated the categories of "defective chimney" and "heating

equipment, defective" as the third and fourth leading causes of fire, with the

"defective chimney" category accounting for 10 per cent of the fires and the

"heating equipment, defective" category accounting for 9.6 per cent of the fires.

Thus, the chimney and heating equipment categories together were identified as

the causes accounting for 19.6 per cent of all the fraternity, sorority and college

dormitory fires in the 1954 data. However, the 1995 data for fraternity and sorority house fires indicated the category of "heating" was the eighth leading

cause of fires accounting for only 5.2 percent of all the fires. While, for the 1995

data concerning college dormitories the "heating" category was the ninth leading

cause of fire accounting for only 2.5 percent of all the fires. It would appear the

decline of the heating related causes of fire including the defective chimneys

between the 1954 and the 1995 data might be related to the decline in the use of

solid fuel heating devices and systems prevalent in the 1940's and 1950's to the

use of the fluid fired heating devices and systems that are prevalent in the 1990's.

It should be noted that all of these comparisons relative to the 1954 NFPA

data and the 1995 data are limited by the different procedures used in the

collection, compilation and analysis of the data, including the different

identification of the fire cause categories. Thus, it might appear to be valid to

relate the "open fireplaces" category in the 1954 data to the "open flame, ember,

torch" category of the 1995 data. It would appear the 1995 category is broader and

could include other sources in addition to "open fireplaces." However, it is

interesting to note the category of "open fireplaces" in the 1954 data was identified

as the eighth leading cause of the fires accounting for 5.3 per cent of all the fires.

While in the 1995 data for the fraternity and sorority houses the category of "open

flame, ember, torch" was the seventh leading cause of fires accounting for 5.4

percent of all the fires. The 1995 data for the college dormitories indicated the category of "open flame, ember, torch" was also the seventh leading cause of fires,

accounting for 5.1 percent of all the fires.

The National fire Protection Association has also prepared a summary of all

the fraternity and sorority house fires reported to them involving occupant

fatalities from March 16, 1975 through May 12, 1996. (16) This summary also

includes the reported property loss involved in the fire when that information was

reported to the FIDO (Fire Information Data Organization) of the National fire

Protection Association. This twenty-one year summary includes fatal fires in

nineteen fraternity house fires and one sorority house fire. This data summary

would tend to provide credibility to the following statement by Isner relative to the

predominance of fatal fraternity house fires in his report on the Fraternity House

fire of May 12, 1996 in Chapel Hill, North Carolina: (10)

> "Since the NFPA began compiling data on fire losses in the early
> 1970s, all of the catastrophic life-loss fires (involving five or more
> deaths) in fraternity or sorority housing have occurred in fraternity
> houses." p. 36.

This NFPA summary includes five of the fraternity house fires that were

included in the six fraternity house fires studied extensively in the first four

sections of the this report with their data presented in Tables 1 through 4.

(9)(10)(12)(14)(21) This National Fire Protection Association 1996 summary of

twenty-one years of fatal fires in fraternity and sorority houses is presented in

Table 8. (16)

Date	Facility & Location	Fat.	Inj.	Dollar Loss
Table		**8**		
3/16/75	Frat., Burlington, VT.	1	1	100,000.00
1/08/76	Frat., Columbus, OH.	2	6	No report
8/29/76	Frat., Baldwin City, KS.	5	2	No report
1/14/78	Frat., University Park, TX.	1	2	525,000.00
4/05/80	Frat., Eugene, OR.	1	1	60,000.00
9/09/82	Frat., Philadelphia, PA.	1	8	No report
5/28/83	Frat., Bridgewater, MA.	1	1	75,000.00
12/11/83	Frat., Austin, TX.	1	1	335,000.00
1/06/84	Frat., Thibodaux, LA.	1	0	No report
4/11/84	Frat., Lexington, VA.	1	0	420,000.00
10/21/84	Frat., Bloomington, IN.	1	30	100,000.00
12/20/84	Frat., Geneseo, NY.	1	0	No report
3/03/85	Frat., San Jose, CA.	1	1	117,000.00
4/19/86	Frat., Danville, KY.	1	0	No report
9/08/90	Frat., Berkeley, CA.	3	2	2,100,000.00
12/08/90	Frat., Erie, PA.	1	4	No report
2/13/92	Frat., California, PA.	1	0	70,000.00
10/24/93	Sor., LaCrosse, WI	1	2	No report
10/21/94	Frat., Bloomsburg, PA.	5	0	70,000.00
5/12/96	Frat., Chapel Hill, NC.	5	3	475,000.00
	Total	**35**	**64**	**4,447,000.00**

As indicated in Table 8, there were a total of 35 occupant fatalities in these nineteen fraternity house fires and one sorority house fire in this twenty-one year period. Three of the fraternity house fires had the maximum of five fatalities in each fire, [Baldwin City, KS. (14); Boomsburg, PA. (21) and Chapel Hill, NC. (10)] one fraternity house fire had three fatalities, [Berkeley, Ca. (9)] and one had two fatalities, [Columbus, Oh. (12)]. The remaining fourteen fraternity houses and the one sorority house fire all resulted in a single occupant fatality.

There were a total of 64 reported occupant injuries in these nineteen fraternity house fires and one sorority house fire. Almost fifty percent of the injuries, involving thirty occupants occurred in the single fraternity house fire in

Bloomington, Indiana on October 21, 1984. The remaining four largest injuries consisted of eight injuries in the fraternity house fire of September 9, 1982 in Philadelphia, Pa., six injuries in the fraternity house fire of January 8, 1976 in Columbus, Ohio and four injuries in the fraternity house fire of December 8, 1990 in Erie, Pa. The fraternity house fire of May 12, 1996 in Chapel, NC (10) in addition to the five fatalities had three injuries. There were three fraternity house fires with two injuries each, and five fraternity house fires with a single injury in addition to the fatalities. There were five fraternity house fires with a single fatality and no injuries and one fraternity house fire with five fatalities and no injuries [Bloomsburg, PA. (21)]. It should be noted, the single sorority house fire of October 24, 1993 in LaCrosse, WI. resulted in one occupant fatality and two occupant injuries. It is apparent from a review of Table 8 that the fire caused life loss problem in fraternity and sorority house fires is definitely a fraternity house fire problem.

Twelve of the nineteen fraternity house fires had an identified and reported dollar loss from the fire incident with the range of reported losses varying from $60,000 in the Eugene, OR. fire of April 5, 1980 to $2,100,000 in the Berkeley, CA. fire of September 8, 1990. The total dollar loss from these twelve fraternity house fires as indicated in Table 8 was $4,447,000.00 and the mean dollar loss for the twelve facilities was $370,583.00.

A review of the National Fire Protection Association *Fire Journal* and *NFPA Journal* "Firewatch" sections as published by the Association in 1995 provided additional listings of college dormitory, fraternity house and sorority house fires. (16) These listing were from the twenty-five year period from 1971 through 1995 and identified eighteen college dormitory fires, five fraternity house fires and one sorority house fire. Two occupant fatalities were identified in these twenty-four fires. One fatality occurred in a single student room fire which occurred in a 13 story fire resistive dormitory. The other occupant fatality occurred in a 3 story frame fraternity house fire reportedly caused by electrical wiring in a second floor wall.

A7. FIRE PREVENTION STRATEGIES-PROGRAMS

It is apparent from a review of the fire cause data previously presented in Tables 5, 6, and 7, indicated the leading causes of fires in both fraternity houses and college dormitories to be the occupant related causes of "incendiary or suspicious" "cooking" and "smoking". Tables 3 and 8 appear to indicate relative to the occurrence of occupant fatalities and injuries that fraternity houses pose the most severe life safety problem for the past two decades. The college dormitory fire problem relative to single fire multiple occupant fatalities and injuries appeared to have peaked in the 1960's and 1970's.

David Breen, while at Harvard University surveyed nine private institutions relative to their fire evacuation drill procedures and frequency in the student residence dormitories. (2) In 1979 at the time of this survey three of the nine universities, one-third of the population conducted no evacuation drills in their college dormitories. One university conducted two drills per year, one at the beginning of each semester. Three universities conducted four drills per year and two universities conducted five drills per year. Of the six universities conducting evacuation drills, three conducted them only in the evening hours and three conducted their drills in both a.m. and p.m. hours. Four of the six institutions conducting evacuation drills included unannounced drills in their procedures. Breen also reported that in 1979 New York State Law required four evacuation drills per year, and in Rhode Island the State Law required two evacuation drills per year.

Nygren (18) has indicated the unique problems involved with a high-rise college dormitory of 28 stories in two wings with 1,100 students in each wing. The dormitory has a suite arrangement as is prevalent with high-rise college dormitories with four suites each with 12 residents per floor for a total of 48 residents per floor in each wing. This style of high-rise housing involved two of the college dormitories (11)(19) with three fatalities, in the analysis of the nine college dormitory fires in Tables 1-4. The evacuation procedures followed in the evacuation drills for this 28 story dormitory involved the residents moving upward and downward within the section of five floors and not to totally evacuate the building, unless directed by fire department personnel over the public address system. Thus, the fire floor of one wing and two floors above and below the fire floor would be evacuated with the activation of the fire alarm system. This is an evacuation procedure followed in many high-rise buildings of residential, commercial and mercantile occupancy.

Following the fraternity house fire in Chapel Hill, North Carolina on May 12, 1996 with five fatalities and three injuries the Chapel Hill Town Council on November 11, 1996 enacted fire prevention ordinances requiring the installation of automatic sprinkler systems. (8)(10) These ordinances require the complete automatic sprinkler protection in all new fraternity or sorority houses, and the retrofitting of complete sprinkler protection in all existing fraternity and sorority houses within five years.

Sactor (20) has reviewed an inspection program for off campus fraternity and sorority houses which was initiated in 1972 between the University of Maryland, the City of College Park and the Prince George County Fire Department. This program utilizes the University's safety personnel in the annual inspection and the first follow-up inspection of the facilities with the City personnel. The University and City personnel issue the Fire Department correction notices on the first inspection and if corrections are not made by the follow-up

inspection the enforcement of the notices is administered by the Fire Department

with fines and even withdrawal of the County Occupancy Permit, thus closing the

facility to student occupancy. For the twenty-two fraternity and sorority houses on

campus a similar inspection program is followed by the University personnel with

enforcement involving the University's Judicial Programs Office and for students

may involve suspensions, fines or community service. For the organization

enforcement may also involve fines, restrictions on activities and even revoking of

the lease by the University.

Evacuation drills are conducted at both on and off campus fraternities and

sororities once a semester under University personnel supervision and failed drills

are repeated. Criteria of failure are less than 100 percent participation and lack of

notification of the fire department. Morris has emphasized this type of sharing of

the education and enforcement efforts in his 1965 article in the following manner:

(15)

"Firesafety in student housing is literally a matter of life and death, and a
responsibility that ought to be shared by the college, the local or state enforcement
agency, and the organization owning or operating a residence. Each should make
certain that its segment of responsibility has been properly carried out." p. 27.
    Isner in the conclusion of his report on the fraternity house fire in Chapel

Hill, North Carolina of May 12, 1996 resulting in 5 fatalities and 3 injuries

emphasizes the importance of fire prevention inspections of fraternity houses, due to the older types of construction, lack of maintenance and occupant behavior that

is inherent in these facilities: (10)

"Fire prevention inspections are key in reducing the fire risk in fraternity and sorority housing. Fire inspectors need to remain diligent with regard to typical inspection items such as fire extinguishers, fire detection and alarm equipment, general housekeeping, building maintenance, and storage of combustible materials. Moreover, fire inspectors need to enforce all requirements of local codes such as the *Life Safety Code* because these codes regulate interior finish, protection of vertical openings, detection and alarm systems, egress system components, etc. Given the difficulty of maintaining good housekeeping practices in some fraternity and sorority houses combined with historical occupant behavior, the installation of state-of-the-art fire detection equipment and fire alarm systems is essential, as a minimum, and the installation of automatic sprinklers is strongly recommended." p. 37.

A8. SUMMARY AND CONCLUSIONS

Table 1 indicated that of the fifteen college dormitory and fraternity house

fires, thirteen or 86 percent occurred in the early morning hours between 1:00 am

and 7:00 am. This table also indicated in confirmation with Tables 5, 6, and 7 that

the leading cause of fire ignition was incendiary fire setting. Table 1 also indicated

this occupant behavior was prevalent with five incidents and four of the incidents

or 80 percent of the incendiary behavior occurred in college dormitories,

(3)(4)(5)(19) with one incident in a fraternity house. (9) Four of these incendiary

fire incidents or 80 percent occurred between the hours of 1:00 a.m. and 4:00 a.m.

(3)(4)(5)(9) The second leading cause of fires for the college dormitory and

fraternity house population in Table 1 was smoking and all four of the fires ignited from smoking materials occurred in fraternity houses. (1)(10)(14)(21) Upholstered

furniture, predominately sofas were the fuel material immediately ignited in five of

the fraternity house fires (1)(9)(12)(14)(21) and two of the college dormitory fires

reviewed in Table 1. (7)(19)

Sactor has indicated the fire safety problems with fraternity and sorority

houses involve the construction of the facilities, the lack of maintenance and the

occupant behavior as follows: (20)

"The ordinary, wood frame, or even balloon construction of many of the
older buildings can contribute to fire and smoke spread. Open   stairways can also
be found. Multiple additions, sometimes unauthorized, create concealed spaces for
fire to develop and may also change exit and egress patterns and may include
combustible interior finishes."

"General poor upkeep results in unrepaired holes in walls and broken
stairwell and bedroom doors which contribute to smoke and heat spread. Fire
protection systems, emergency lighting, and exit signs when in place can be in
disrepair or out of service."

"Large social events and rough treatment of the facilities causes damage or
excessive wear and tear. Poor fire safety awareness of occupants may result in
blocked exits, flammable decorations, unsafe use of combustibles and smoking
materials, and tampering with fire protection equipment." p. 18.

Table 2 involving the construction variables in the nine college dormitories

and the six fraternity houses emphasizes the difference in the ages of the structures

between the college dormitories and the fraternity houses. Seven college

dormitories were involved with reporting the age of the structure with a range of

39 years to 9 months at the time of the fire with a mean age of 15.5 years. While

the four reporting fraternity houses had an age range from 33 years to 70 years with a mean age of 54.7 years. Thus, when comparing the means ages of the fraternity houses and the college dormitories, the fraternity houses appear to be more than three times as old as the college dormitories. As indicated by Sactor (20) there is also a difference in construction as well as age for the fraternity houses. Table 2 indicated that eight of the nine college dormitories were of fire resistive construction and one was of protected noncombustible construction. (6) While one of the fraternity houses was of protected noncombustible construction, (1) three were of ordinary construction (10)(12)(14) and two were of frame construction. (9)(21) Table 2 also indicated that four of the six fraternity houses had open stairs and the two with enclosed stairs had open doors.

Sactor has identified occupant behavior and organizational differences which may explain the lack of a sorority house in the Tables 1-4 study population and the inclusion of only one sorority house fire compared to nineteen fraternity house fires in Table 8, as follows: (20)

"There are differences between fraternities and sororities which effect the condition of the facilities. Based on experience at UMCP, sororities often have stronger ties to their nationals and closer involvement of house corporations than fraternities. Sororities always have house directors and cleaning services and do not have parties which include alcohol in their facilities." p. 20

From a review of the relevant published literature and the fire record for fraternity, sorority houses and college dormitories from the years of 1944 through

1954, (17) and 1990 through 1996 (16) the following conclusions appear

reasonable:

1. The primary causes of fraternity House and college dormitory fires

appear to be the occupant behavior activities of incendiary fire setting, cooking

and smoking.

1. A. It would appear occupant educational and enforcement procedures

should be focused on the alleviation of these primary fire causes.

1. B. The current educational activities in these occupancies appear to be

generally limited to practice evacuation drills. The educational activities should be

extended to orientation sessions with appropriate visual aids relative to fire ignition

and fire/smoke propagation variables affecting life safety for the residents and

resident assistants in the dormitories and the house directors and the members in

the fraternities and sororities.

2. The Greek organizations fire life loss problem currently appears to be a

fraternity fire problem, often preceded by social events.

3. Fire prevention inspections of college dormitories, fraternity

and sorority houses should be thorough with the adoption and enforcement of

current codes and standards. The enforcement procedures should impact on the

residents and the organizations for continued uncorrected facility features or unsafe

occupant behavior. Such behavior should include false alarms, tampering with fire protection equipment (including detectors and extinguishers), and open flame ignition behavior.

3. Procedures should be initiated to regulate the inclusion of new highly combustible upholstered furniture into both dormitories and fraternity, sorority houses.

4. Procedures should be initiated to provide for the installation of smoke detectors in students rooms and automatic sprinkler systems throughout new fraternity, sorority houses and college dormitories.

When existing facilities are renovated these fire protection systems should be installed.

4. A. The national fraternal organizations should be more involved in providing for the financing of fire protection systems during the renovation of fraternity or sorority houses.

4. B. The national fraternal organizations should be more involved in providing for education and enforcement activites to reduce the occupant behavior involving excessive use of alcohol at the social events which apparently preceded the fire incidents in numerous cases.

A9. SELECTED BIBLIOGRAPHY

1. Best, Richard, "Fire in a Fraternity House Amherst, Massachusetts," *Fire Journal,* 69, 6, (November 1975), 30-33, 105.

2. Breen, David E., *Fire Evacuation Drills at Other universities.* Cambridge, MA: University Health Services, Harvard University, March 21, 1979.

3. Bryan, J. L., J. A. Milke and P. J. DiNenno, *An Examination and Analysis of the Dynamics of the Human Behavior in the Fire Incident at Thurston Hall on April 19, 1979.* College Park, MD: Department of Fire Protection Engineering, University of Maryland, July 31, 1979.

4. Bryan, John L. and James A. Milke, *An Examination and Analysis of the Dynamics of the Human Behavior in the Fire Incident at Chesapeake Hall on February 3, 1980.* College Park, MD: Department of Fire Protection Engineering, University of Maryland, June 30, 1980.

5. Carpenter, Daniel J., Jr., *College Dormitory Fires in Dover, Delaware and Farmville, Virginia.* Washington, D.C.: U.S. Fire Administration, Federal Emergency Management Agency, 1987.

6. Demers, David P. "Ten Students Die in Providence College Dormitory Fire," *Fire Journal,* 72, 4, (July 1978), 59-62, 103.

7. Gaudet, Robert E., "Dormitory Fire Kills Nine, *"Fire Journal,* 61, 4, (July 1967), 5-9.

8. Graham, D'Arcy L., "Getting the Message Out and Getting Things Done," *Sprinkler Age,* 16, 8, (August 1997), 10-13.

9. Isner, Michael S., *Fraternity House Fire, Berkeley, California, September 8, 1990.* Quincy, MA: National Fire Protection Association, Fire Investigation Report, 1990.

10. Isner, Michael S., *Fraternity House Fire, Chapel Hill, North Carolina, May 12, 1996.* Quincy, MA: National Fire Protection Association, Fire Investigation Report, 1996.

11. Lathrop, James K., "Nineteenth-Floor Dormitory Fire Kills One   Student," *Fire Journal,* 70,3, (March 1976), 77-79.

12. Lathrop, James K., "Two Die in Fraternity House Fire," *Fire Journal,* 70, 5, (September 1976), 5-7, 66.

13. Lathrop, James K., "Dormitory Fire Leaves One Dead Twenty-three Hospitalized, Saratoga Springs, New York," *Fire Journal,* 70, 6, (November 1976), 5-7, 13.

14. Lathrop, James K., "Five Die in Fraternity House Fire Baldwin City, Kansas," *Fire Journal,* 71, 3, (May 1977), 21-22, 109-110.

15. Morris, John, "Off-Campus Housing-A college Fire Problem," *Fire Journal,* 59, 6, (November 1965), 26-27.

16. National Fire Protection Association, *Special Data Information Package Dormitories and Fraternity and Sorority Houses.* Quincy, MA: Fire Analysis and Research Division, 1995.

17. National Fire Protection Association, *Occupancy Fire Record College Dormitories, Fraternities and Sororities.* Quincy, MA: FR 54-8, 1954.

18. Nygren, Ronald G., "Alarm Signaling and Evacuation in a High-Rise University Residence Hall, " *Fire Journal,* 66, 2, (March 1972), 5-6, 11.

19. Peterson, Carl E., "Ohio State University Fires, Morrill Tower and Lincoln Tower," *Fire Journal,* 62, 6, (November 1968), 19-23.

20. Sactor, Alan, Fire Inspection Program for Greek Facilities at the University of Maryland at College Park, A Model of University and Community Cooperation. Monograph No. 58, paper presented at 42nd. Annual International Conference on Campus Safety, University of Hawaii at Manoa, June 9-14, 1995.

21. Trembly, Kenneth J., "Catastrophic Fires of 1994," *NFPA Journal,* 89, 5, (September/October 1995), 60.

22. Wolf, Alisa, "Fraternity Fire Kills Five," *NFPA Journal,* 90, 5, (September/October 1996), 61-64.

www.ingramcontent.com/pod-product-compliance
Lightning Source LLC
Chambersburg PA
CBHW081224170526
45165CB00009B/2943